LOMOND AREA

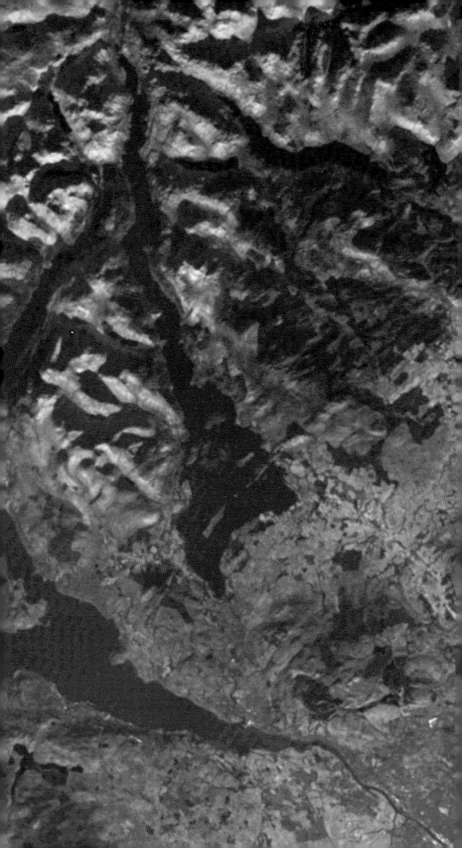

CLASSIC LANDFORMS OF THE
LOCH LOMOND AREA

DAVID J.A. EVANS
University of Glasgow

JIM ROSE
University of London

Series editors
Christopher Green, Michael Naish
and Sally Naish

Published by the Geographical Association
in conjunction with the
British Geomorphological Research Group

THE BRITISH GEOMORPHOLOGICAL RESEARCH GROUP

PREFACE

Geomorphologists study landforms and the processes that create and modify them. The results of their work, published as they invariably are in specialist journals, usually remain inaccessible to the general public. We should like to put that right. Scattered across the landscapes of England, Wales, Scotland and Ireland there are many beautiful and striking landforms which delight the eye of the general public and are also visited by educational parties from schools, colleges and universities. Our aim in producing this series of guides is to make modern explanations of these classic landforms available to all, in a style and format that will be easy to use in the field. We hope that an informed understanding of the origins of the features will help the visitor to enjoy the landscape all the more.

Encouraged by the success of the first edition of the Classic Landform Guides we are pleased to introduce this new series, enhanced by colour photos, new illustrations and with the valuable addition of 1:50,000 map extracts by kind permission of the Education Team, Ordnance Survey. The relevant maps for the area covered in this book are Ordnance Survey 1:50,000 Landranger sheets 56, 57 and 64. Please refer to the current Ordnance Survey Index for details of the relevant 1:25,000 sheets.

Christopher Green *Royal Holloway, University of London*
Michael Naish and Sally Naish *Hayes, Kent*

© D.J.A. Evans and J. Rose, 2003

This book is copyright under the Berne Convention. All rights are reserved. Apart from any fair dealing for the purpose of private study, research, criticism or review as permitted under the Copyright, Designs and Patents Act 1988, no part of this publication may be reproduced, stored in a retrieval system, or transmitted in any form or by any means, electronic, electrical, chemical, mechanical, optical, photocopying, recording or otherwise, without the prior written permission of the copyright owner. Enquiries should be addressed to the Geographical Association. The author has licensed the Geographical Association to allow, as a benefit of membership, GA members to reproduce material for their own internal school/departmental use, provided that the copyright is held by the author. This waiver does not apply to Ordnance Survey maps, questions about which should be referred to the Ordnance Survey.

ISBN 1 84377 072 5

This edition first published 2003

Published by the Geographical Association, 160 Solly Street, Sheffield S1 4BF. The views expressed in this publication are those of the author and do not necessarily represent those of the Geographical Association. The Geographical Association is registered charity: no 313129.

CONTENTS

Introduction	*6*
The Campsie Fells	*21*
Conic Hill and the islands of South Loch Lomond	*27*
Drumbeg and Gartness	*31*
Croftamie	*38*
Cameron Muir, Carnock Burn and Finnich Glen	*40*
The Whangie	*42*
The Endrick Water floodplain	*44*
Rowardennan Pier	*46*
Western Forth Valley	*48*
Glossary	*54*
Bibliography	*56*

Cover photograph: View northwards across Loch Lomond and the neighbouring highlands taken from Duncryne Hill. Ben Lomond is the highest snow-covered peak to the right of the Loch. Conic Hill and the island Inchcailloch, marking the location of the Highland Boundary Fault, trend across the middle distance. *Photo:* D.J.A. Evans.

Frontispiece: Satellite image of the Loch Lomond and upper Forth Valley region, showing the clear break in topography on either side of the Highland Boundary Fault. *Source:* British Geological Survey (www.bgs.ac.uk).
© NERC. All rights reserved.

Acknowledgements

Maps reproduced from Ordnance Survey 1:50,000 Landranger mapping, digital map data © Crown copyright 2003. All rights reserved.
Licence number: NC/03/10001/7849.

Copy editing: Kath Davies
Illustrations: Paul Coles
Series design concept: Quarto Design, Huddersfield
Design and typesetting: Arkima Ltd, Dewsbury
Printing and binding: EspaceGrafic, Spain

INTRODUCTION

The geology and associated landforms of the south Loch Lomond basin are dominated by the Highland Boundary Fault. The fault separates the distinctive rock types of the lowland block to the south and the highland block to the north (Figure 1 and frontispiece) and has had a significant impact on not only the long term development of the relief but also the weather, vegetation, wildlife and cultural history of the region.

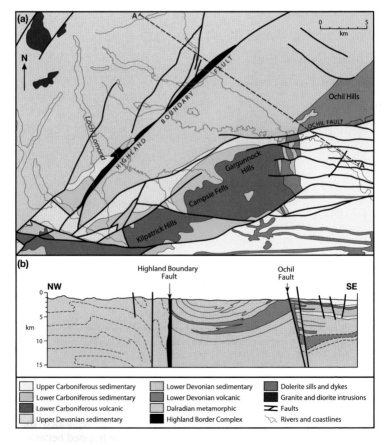

Figure 1: The Loch Lomond basin and the western Forth Valley: (a) geology of the region, and (b) cross-section of the geology along the transect line in (a), showing major rock types and ages. After Browne and Mendum (n.d.).

Table 1: The main events relating to the formation of the rocks in the Loch Lomond area.

Age (MYBP)	Period and palaeolatitude	Event
300		
	0°N CARBONIFEROUS 5°S	Mudstones and limestones deposited in tropical shallow seas and coastal plains - coal forms from decaying rainforests. Volcanoes erupt basalt lavas and ashes.
360		
	DEVONIAN 15°S	Earth movements produce folding, faulting, uplift and erosion. Large rivers deposit sandstones and conglomerates. Volcanoes produce lavas and debris flows
410		
	SILURIAN	Emplacement of Highland Border Complex. Highland and Midland Valley crustal blocks joined
435		
	22°S ORDOVICIAN	Main Caledonian Mountain building
495		
	CAMBRIAN	Deposition and formation of Highland Border Complex components
570		
600	PRECAMBRIAN	Start of Caledonian mountain building

Note: MYBP – million years before present – standardised at 1950.

The two geological blocks came together between 450 and 420 MYBP (Tables 1 and 2) but the rocks of the Highlands are considerably older than the rocks that now occupy the Lowlands. The latter were deposited during the Devonian and Carboniferous periods, after the two geological blocks came together. The Highlands are resistant upland massifs composed of deformed and metamorphosed Precambrian rocks, initially uplifted during the Caledonian mountain building episode. Associated with the Highland Boundary Fault is a narrow band of rocks called the Highland Border Complex. These are mainly marine rocks of Cambrian and Ordovician age that were deposited in the vast Iapetus Ocean that originally separated the Highland and Lowland blocks (Figure 2). They were trapped between colliding continents, caught between the two blocks as they came together and compressed and deformed as the blocks slipped past each other.

Table 2: The main events in the shaping of the relief of the Loch Lomond area.

Age (years BP)	Period and palaeolatitude	Event
200		Major clearance of peat in Forth Valley
500		Clearance of forests
5000		
6500		Loch Lomond is a sea loch
6800		Carse clays deposited in Forth Valley
10,000		End of Loch Lomond Stadial/Readvance and Flandrian Interglacial (Holocene) begins
11,000		Start of Loch Lomond Stadial/Readvance
	QUATERNARY	Windermere (Lateglacial) Interstadial
12,800		Dimlington Stadial ends
13,000		Loch Lomond is a sea loch until 11,000 years BP
27,000		Build up of Dimlington Stadial ice sheet on Rannoch Moor
		Numerous glaciations and interglacials
		- large glacial erosional features (e.g. Loch Lomond) are eroded
	56°N	- Tertiary drainage is modified
2 million		Global climate cools and initiates the glacial/interglacial cycles of the Quaternary
	48°N	Present-day drainage patterns initiated
	TERTIARY	Uplift and erosion
65 million		

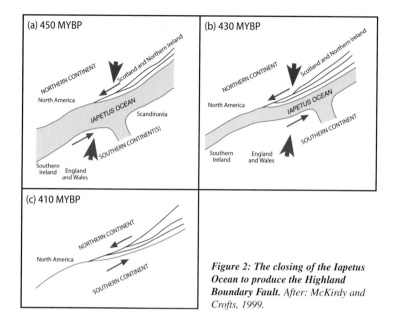

Figure 2: The closing of the Iapetus Ocean to produce the Highland Boundary Fault. After: McKirdy and Crofts, 1999.

Little is known about the long-term or pre-Quaternary evolution of Loch Lomond and the surrounding terrain. The Loch is Scotland's largest freshwater lake at 33km long and 7.5km across at its widest point and it cuts across the regional geological structure. This is thought to reflect excavation and overdeepening of preglacial river valleys by glacial erosion. The preglacial drainage pattern, dating to the Tertiary (Table 2), was very different from that of today (Figure 3). Three drainage systems are thought to have drained runoff from the Loch Lomond area towards the east:

1. the Glen Falloch/Glen Dochart system draining north-east to the Tay;
2. the Loch Arklet/Loch Katrine system draining to the Teith; and
3. the Glen Luss/south Loch Lomond system draining to the Forth.

The interfluves separating the three major drainage basins were breached by glacier ice during the Quaternary, producing the present drainage pattern. The impact of glacial erosion is evident in the large number of truncated spurs that occur on both sides of the northern part of the Loch where it possesses a classical U-shaped cross profile (Figure 4). These features were the result of erosion by the outlet glaciers of the Scottish sector of the British ice sheet centred on Rannoch Moor. After breaking through the bedrock ridges of the Highland Boundary Fault, glaciers probably extended an erosional trough northwards towards the ice dispersal centre, thereby dissecting the original west to east trending river valleys.

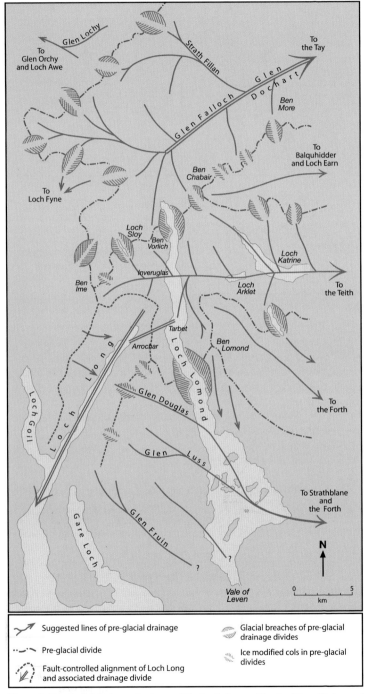

Figure 3: Preglacial drainage features of the Loch Lomond basin. After: Linton and Moisley, 1960.

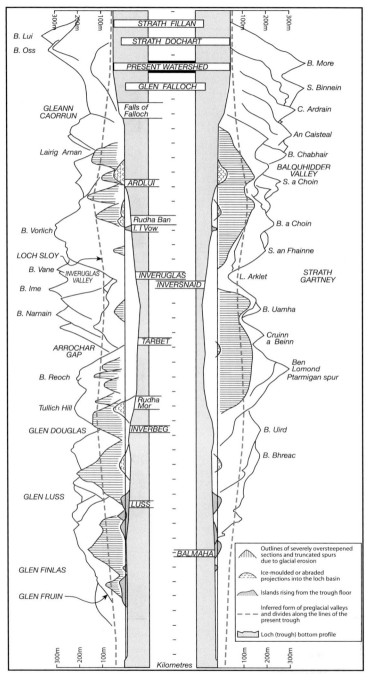

Figure 4: Cross profiles along both shores of Loch Lomond, showing the loch (trough) bottom profile and outlines of major summits to the east and west. Note: The vertical distance between the trough bottom profile and the broken line represents probable depth of rock eroded by glaciers. After: Linton and Moisley, 1960.

— CLASSIC LANDFORMS —

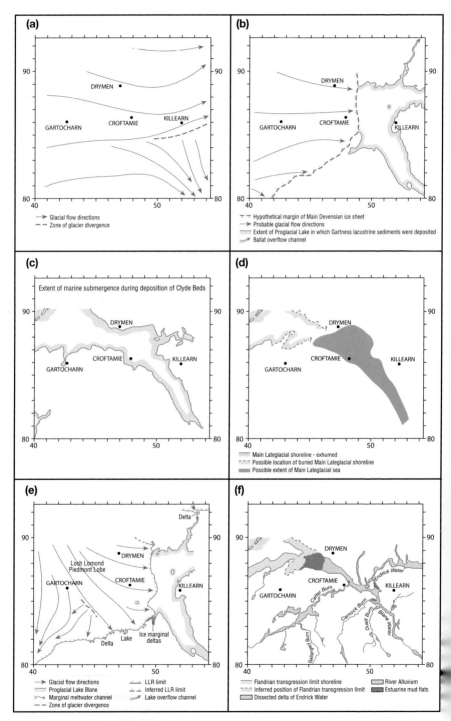

Figure 5: Glacial effects on the area covered by this guide.

The area contains a diverse selection of landforms and sediments relating to the last glaciation (the Dimlington Stadial of the Late Devensian) and the following period leading up to the present, Flandrian Interglacial (Tables 2 and 3). These sediments and landforms record the advance and recession of glacier ice in association with proglacial lake damming and changing sea levels. From this evidence the following events have been reconstructed (Table 3; Figures 5 and 6).

During the later stages of the Dimlington Stadial glaciation, the regional ice sheet inundated the region and deposited the Wilderness Till. Drumlin orientations, **till fabrics** and **erratics** show that the ice flowed west to east across Dumbarton Muir and the south side of Loch Lomond as far as the Campsie Fells where its basal flow was diverted north-eastwards across the Clyde-Forth watershed and south-eastwards along the Blane Valley. Basal ice flow diverged over the present location of Killearn (Figure 5a). During the recession of ice from the area, proglacial lakes were dammed temporarily by the ice margin in the Blane and Endrick valleys (Figure 5b), into which the Gartness laminated silts and clays were deposited. Lake water drained north-eastwards towards the North Sea via the spillway gorge at Ballat.

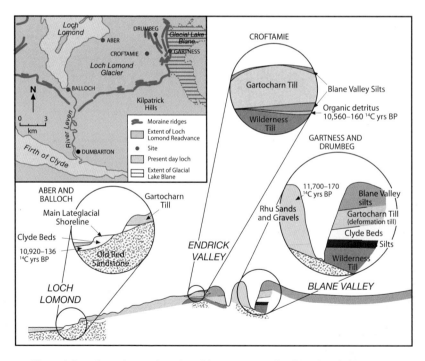

Figure 6: Locations of general stratigraphic sequences and radiocarbon dates critical to reconstructions of glacial and sea level events during the Dimlington and Loch Lomond Stadials and the intervening Windermere Interstadial. After: Rose, 1989.

Table 3: Summary of events, stratigraphic units and depositional environments around the Loch Lomond basin and Clyde Estuary from the Late Devensian glaciation to the present.

Event and radiocarbon	Sediments/ stratigraphic unit	Processes/palaeogeography age unit
Flandrian Interglacial (10,000 YBP to present)	Beach sediments, shoreline landforms and associated river terraces	b) Marine limit of 14m OD inside Loch Lomond Readvance moraine. Marine invasion of Loch Lomond 6900-5400 YBP. Deltas in Endrick and Fruin valleys. a) Relatively low sea level.
Loch Lomond Readvance (Stadial) (11-10,000 YBP)	Blane Valley laminated silts and clays	a) Proglacial lake sediments in Endrick/Blane valleys with drainage into North Sea via Firth of Forth.
	Rhu glacifluvial sands and gravels	b) Glacifluvial sediments of Loch Lomond Readvance with associated moraine ridges, eskers, kames and outwash terraces and deltas. a) Relatively low sea level (<2.3m at Rhu).
	Gartocharn Till	a) Loch Lomond Readvance till. Usually shelly due to derivation from reworked Clyde Beds. Often deformed (glacier snout oscillations recorded in glacitectonic structures at Drumbeg). Associated with radial flowing piedmont glacier lobe (Loch Lomond Glacier) that constructed streamlined hills and Drumlins. Drumlins are superimposed on larger Dimlington Stadial streamlined hills.
	Main Lateglacial shoreline	a) Large shore platform and rock cliff associated with relatively low sea level (ca. 10m at Ardmore).
Windermere Interstadial (12,800-11,000 YBP)	Clyde Beds	a) Estuarine silts associated with falling sea level. Rich arctic-boreal marine fauna.
	Gartness gravel, sands and laminated silts and clay	a) Proglacial lake sediments in Endrick Valley. Lake dammed by westward receding margin of Dimlington Stadial glacier.
Dimlington Stadial (Late Devensian) glaciation	Wilderness Till	c) Westward recession across region and down Clyde Estuary. b) Marine sediments deposited in Clyde Estuary from floating glacier margin. a) North-west to south-east glacier flow across Dunbartonshire hills and Clyde Estuary. West to east flow across south side of Loch Lomond – origin of main drumlin field.

After deglaciation of the Loch Lomond basin the sea inundated the Endrick and Blane valleys up to an altitude of approximately 60m OD (Ordnance Datum – standardised sea level in Great Britain), temporarily turning Loch Lomond into a sea loch (Figure 5c). Such marine inundation is characteristic of recently deglaciated coastlines because the crust is still 'rebounding' after being depressed by the weight of the ice sheet (glacioisostatic rebound). The crustal depression created by ice sheet loading is approximately 0.3 times the ice thickness (e.g. an ice sheet 1km thick will depress the crust beneath it by a maximum of 333m). Because it requires several thousand years for the crust to rebound from an ice load after deglaciation, the sea may flood into low lying areas once the ice has receded. The level to which the sea rises is known as the marine limit and is demarcated by shoreline features such as beaches, marine platforms and deltas. In the Endrick and Blane valleys marine inundation is recorded by the Clyde Beds, which are estuarine silts containing marine fauna of arctic and boreal affinities. Shells from the Clyde Beds have been radiocarbon dated to 12,700-11,500 years BP (BP – before present, standardised at 1950). This indicates that marine inundation in the Loch Lomond basin occurred during the Lateglacial or Windermere Interstadial, a period of relative warmth immediately following the Dimlington Stadial and named after the type site at Lake Windermere in the English Lake District.

Throughout Britain, beyond the margins of the Loch Lomond Readvance glaciers a clear stratigraphic and ecological record of the Windermere Interstadial, the Loch Lomond Stadial and the early Flandrian Interglacial accumulated in small lake basins (Figure 7). This record, which has been radiocarbon dated in many places,

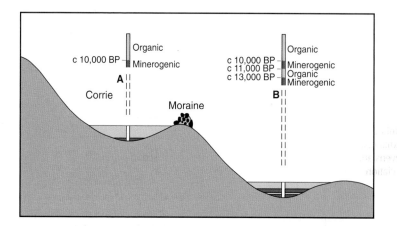

Figure 7: Typical peat bog stratigraphies located: (a) inside, and (b) outside a Loch Lomond Stadial moraine, with associated radiocarbon dates.

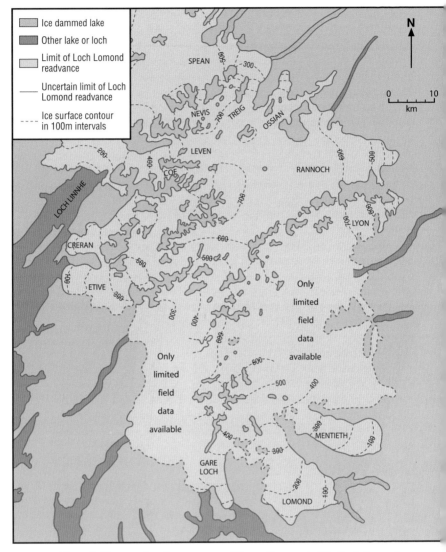

Figure 8: Extent and surface contours of the southern part of the Loch Lomond Stadial West Highland Glacier Complex. After: Thorp, 1991.

consists of two sediment couplets each comprising inorganic material (lake clays) overlain by an organic lake deposit called **gyttja.** The lower lake clays were deposited by surface runoff carrying sediment from freshly deglaciated and sparsely vegetated slopes. This is overlain by the gyttja as drainage into the lakes began to transport plant remains from the vegetation cover developing during the Windermere Interstadiial. The upper lake clays indicate renewed deterioration of climate and of vegetation cover during the Loch

Lomond Stadial. The final capping of gyttja signals a return to warmer conditions at the beginning of the Flandrian Interglacial. Thus **the** occurrence of two inorganic lake clay layers in the sediments of an upland lake or pond indicates that the site lay beyond the limits of the Loch Lomond Readvance.

Relative sea level dropped throughout the Windermere Interstadial as a result of crustal rebound. This continued into the next phase of landscape change in the area, the Loch Lomond Stadial. Recognised throughout Europe and increasingly around the world, this was a period of renewed cold (Younger Dryas) following the Windermere Interstadial. As such it interrupted the initial climate warming of the present (Flandrian) interglacial, giving rise to the renewed growth of icefields and **cirque** glaciers in upland Britain. The evidence for the Younger Dryas in the Loch Lomond basin has prompted the use of the area as the British type site for the cold episode of mountain glaciation and hence the terms Loch Lomond Stadial and Loch Lomond Readvance, the latter referring specifically to the activity of glaciers during the Stadial. The largest outlet glaciers in Britain at this time were nourished by the West Highland glacier complex centred over Rannoch Moor (e.g. Loch Lomond Glacier and Menteith Glacier; Figure 8). Although sea level was falling throughout the Loch Lomond Stadial, it was at this time that the Main Lateglacial Shoreline formed. This is a prominent marine platform and associated cliff cut by a combination of **periglacial** processes and intense wave activity due to increased storminess (Figure 5d). This shoreline is locally buried by the Gartocharn Till which was deposited by the Loch Lomond Glacier when it advanced into the south Loch Lomond basin. This till contains numerous fragments of marine shell derived from the Clyde Beds.

The limit of the Loch Lomond Readvance is marked by a prominent **moraine** ridge. Although several recessional moraines occur just inside the outermost limit of the readvance, most of the terrain covered by the Loch Lomond Glacier is characterised by a hummocky till cover and **kames** draped over drumlins (Figure 9). As the glacier flowed into the lower part of the Endrick Valley it once again dammed the natural drainage, producing proglacial Lake Blane (Figure 5e). Lake waters again escaped along the Ballat spillway where they were ponded for a short period against the southern margin of the Menteith Glacier, as recorded by a small delta on the south side of the Loch Lomond Readvance maximum moraine just north of Garrauld. The Blane Valley laminated silts were deposited in Lake Blane during this episode.

Changes in the south Loch Lomond basin during the present Flandrian Interglacial have been dominated by fluvial responses to sea level change. Between approximately 10,500 and 6900 years BP, Loch

Lomond was a freshwater lake with a level controlled by the Loch Lomond Readvance moraine barrier in the Vale of Leven. The sea again inundated the loch between 6900 and 5450 years BP, rising to an altitude of 12m OD as a result of sea level rise caused by the melting of the North American and Scandinavian ice sheets (**glacioeustatic effect**). At this time the upper Forth Valley was also flooded, giving rise to the deposition of estuarine sediments (carselands) as far west as Flanders Moss and the production of the Main Postglacial shoreline in the region. Marine erosion at this stage resulted in the localised exposure of the Main Lateglacial shoreline from beneath its cover of

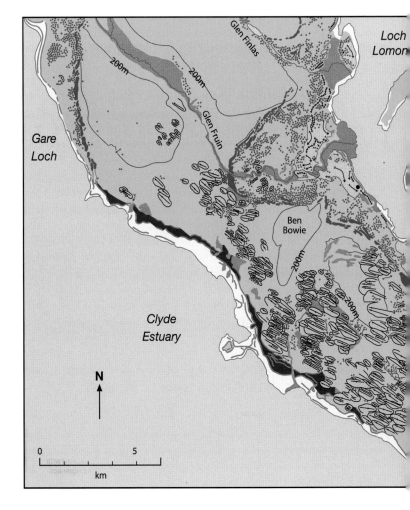

Figure 9: Quaternary geomorphology map of the southern Loch Lomond basin and north east Clyde Estuary. After: Rose, 1980, 1981.

glacial sediments and the formation of the Endrick raised delta (Figure 5f). Continued glacioisostatic rebound of the region eventually overcame this short pulse of global sea level rise, thereby resulting in a renewed fall of sea level from approximately 5000 years BP until the historical period.

The following chapters describe specific geomorphological features in southern Loch Lomond and the western Forth Valley. Figure 10 indicates the location of and access to places and features mentioned in the text.

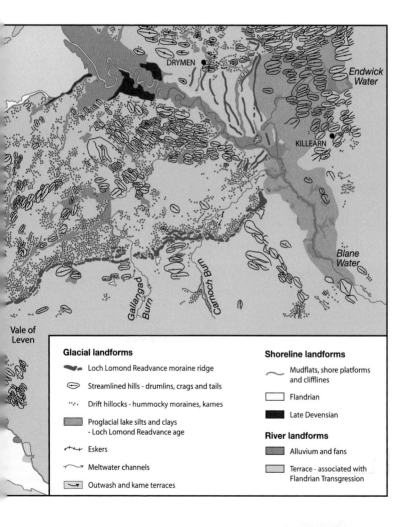

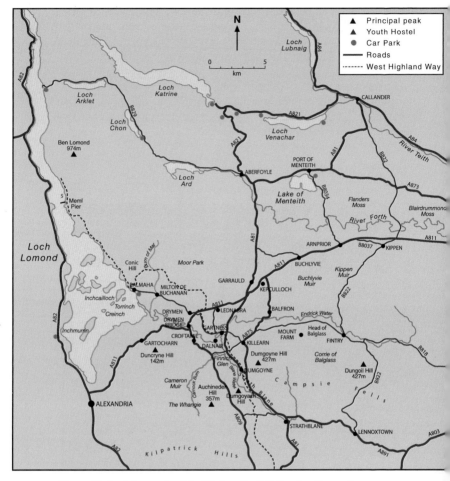

Figure 10: Loch Lomond and the Western Forth Valley: location and access.

THE CAMPSIE FELLS

Photo 1: The Campsie Fells *and the volcanic plug of Dumgoyne Hill viewed from the Kilpatrick Hills. Photo: D.J.A. Evans.*

The upland plateaux of the Campsie Fells and Kilpatrick Hills dominate the northward view from the outskirts of Glasgow. Their tabular appearance with tiered margins and immediately recognisable cone-shaped outliers like Dumgoyne Hill (Photo 1) reflect a violent geological origin. The sub-horizontal strata responsible for the tabular shape and tiered margins of the Campsie Fells are a product of numerous lava flows (the Clyde Plateau Lavas) deposited during the Lower Carboniferous. The cone-shaped outliers (e.g. Dumgoyne (NS 542828), Dumfoyne, Duncryne Hill and Dumgoyach Hill) are the remains of **volcanic plugs**, the former vents from which the Clyde Plateau Lavas spilled out during successive eruptions. The Kilpatrick Hills and Campsie Fells are the remnants of a once more extensive lava plateau which has been dissected by river and glacial erosion during the Quaternary Period. Indeed, the removal of a considerable thickness of the lava sheets by these geomorphological processes during the production of the Blane Valley has exposed the deeper volcanic plugs of Dumgoyne, Dumfoyne (both basalt agglomerates) and Dumgoyach Hill (basalt). Glacial erosion of the lavas directly south of Loch Lomond during successive advances of the highland glaciers down the Loch Lomond basin has exposed the plug of Duncryne Hill in a similar way.

Although ice sheet glaciation inundated the landscape, the erosional impact of glaciation on the upland lava plateau of the Campsie Fells is manifest mostly in the spectacular **cirques** on the north facing flank,

Photo 2: The northern face of the Campsie Fells viewed from the north east, showing the Coire of Balglass on the left. Photo: D.J.A. Evans.

for example the Corrie of Balglass (Photo 2). The corrie was last occupied by a glacier during the Loch Lomond Stadial when an end moraine was constructed on the outer floor of the basin. A small cliff exposure near the Head of Balglass, outside the Loch Lomond Stadial end moraine, contains a sedimentary sequence comprising tills and glacial lake sediments. The tills are regional glacial sediments deposited by highland glacier ice, as indicated by erratics derived from the highland region. The lake sediments were deposited in lake waters dammed between the Campsie Fells and the regional ice during ice sheet deglaciation.

Following deglaciation, landscapes tend to undergo a period of adjustment to the new non-glacial conditions. This period is referred to as the paraglacial cycle and is characterised by an initial phase of rapid adjustment marked by very active slope and river processes which gradually settle down to the level of geomorphic activity that we see today (Figure 11). The very active phase reflects the instability of the unconsolidated glacial sediments and of oversteepened rock slopes which have lost the lateral support of glacier ice. The decline in geomorphic activity through time reflects the stabilisation of slopes. Landslides often occur in recently deglaciated terrain as a process in the rapid and large-scale adjustment of oversteepened slopes. The numerous landslides around the margins of the Campsie Fells possibly record this type of paraglacial adjustment during postglacial times, although the triggering of landslides by earthquake activity cannot be dismissed.

The landslides of the Campsie Fells are predominantly rock avalanches and debris or rotational slides (Figure 12). The north-facing scarp of the Campsie Fells is widely disrupted by rotational landslides. These are recognisable by tensional cracks and downslope displacement of blocks near the source wall and by compressional ridges at the foot of the landslide. The tensional cracks have been produced by sliding of lava blocks downslope from the steep scarp face, whereas the compressional ridges result from the piling up of individual blocks in the lower part of the landslide.

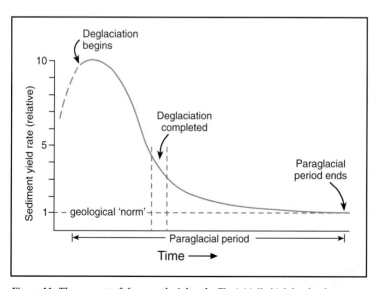

Figure 11: The concept of the paraglacial cycle. The initially high levels of geomorphic activity (represented by sediment yield or production) in freshly deglaciated upland drainage basins and the gradual reduction in geomorphic activity as the landscape settles down to the geological 'norm' of the present interglacial.

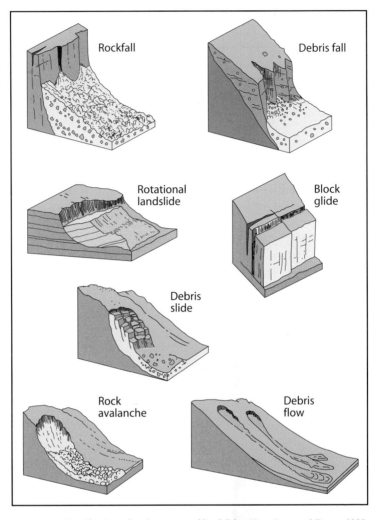

Figure 12: Classification of various types of landslide. After: Benn and Evans, 1998.

Impressive relict rock avalanches occur above the villages of Strathblane and Lennoxtown on the west and south scarps of the Campsie Fells respectively (Figure 10, page 20). In all cases the avalanche scars or source areas are recognisable as steep and unvegetated rock walls in an otherwise degraded cliff. The large volume of rock that has failed catastrophically from these source areas now lies downslope on the gentle footslopes, forming a series of wave-like hummocky ridges which can be mistaken for the moraines of a small glacier (Photo 3). In rock avalanches the falling mass of blocks may act like a fluid and flow over alarmingly great distances, sometimes travelling across valley bottoms and up the opposite slope.

Regardless of the exact mode of landsliding, some geological structures and physiographic settings favour large scale failure of bedrock. Several of these structures and settings are common in the Campsie Fells, making it unsurprising that landslides are ubiquitous in the area. For example, strong rocks overlying weaker rocks can fail along deep-seated slide planes. Additionally, alternating strata possessing different rock strengths are prone to failure due to deformation, weathering and high porewater pressures in the weaker beds. The interbedded lavas and associated intrusions of the Campsie Fells are very likely to be susceptible to such failure. Finally, tectonic activity within a region may have fractured and deformed the rocks and produced densely spaced fracture planes. If slopes in such rocks are oversteepened, for example by glacial erosion or fluvial undercutting, they may become prone to catastrophic failure. This is particularly common in upland terrains where bedrock undergoes pressure release after the removal of the lateral support of a glacier.

Photo 3: A large rock avalanche on the southern face of the Campsie Fells above Lennoxtown. Photo: D.J.A. Evans.

Access

The Campsie Fells can be accessed by foot from various directions but a full appreciation of the Dumgoyne and Dumfoyne volcanic plugs is afforded by a short climb along a well-trodden path from the Glengoyne distillery on the A81 (NS 527826). Dumgoyach Hill is heavily wooded but is an impressive site rising from the flat terrain of the Blane Valley, best observed from the A875 or from the West Highland Way which skirts the base of the hill. A marvellous short walk, providing excellent views of the Loch Lomond basin, climbs Duncryne Hill from a small lay-by on a country road leading south out of Gartocharn (NS 433856).

The Corrie of Balglass is on private land but access is possible from the end of the road leading to a farm called Mount (NS 582868), where permission should be sought before venturing on to the fells.

The Strathblane rock avalanche is best viewed from the B821 at NS 545798, and the Kippen slumps look best in late evening sunlight from Kippen Muir on the B822. Indeed, numerous landslides can be viewed along this road and the B818 between Kippen and Balfron via Fintry.

The Lennoxtown landslides are best viewed from the A891 and can be visited by using the road that crosses the features, the B822 to Fintry.

CONIC HILL AND THE ISLANDS OF SOUTH LOCH LOMOND

Although its name implies a cone shape, Conic Hill (NS 433924) is in fact a **hogsback ridge** and its name probably originates from the Gaelic term *A'Coinneach* (the moss or bog). The summit comprises several parallel ridges that represent the upstanding edges of individual strata of the Devonian (Old Red Sandstone) sandstones and conglomerates and rocks of the Highland Boundary Complex (Figure 1, page 6). On the Balmaha side of the hill the rocks can be seen to dip very steeply towards the south. This striking disturbance of the lowland strata was produced by activation of the near vertical Highland Boundary Fault, which lies directly to the north. The view from the summit of Conic Hill towards Loch Lomond reveals a chain of islands (Inchcailloch, Torrinch, Creinch and Inchmurrin; Photo 4a), whose alignment and elongated forms are a product of movement along the Highland Boundary Fault. The ridged summit of the hill reflects the underlying geological structure, specifically the near vertically inclined strata of the Old Red Sandstones and the Highland Boundary Complex (Photo 4b). To the south of this line lie the lowland sedimentary rocks of Devonian age and to the north lie the more resistant Precambrian age metamorphic rocks that form the mountainous highlands.

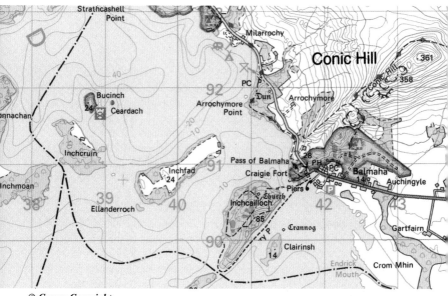

© *Crown Copyright*

Photo 4: (a) View from the summit of Conic Hill looking towards the islands of southern Loch Lomond, and (b) the ridged summit of Conic Hill viewed from the B837. Photos: D.J.A. Evans.

The lower south-facing slopes at the east end of Conic Hill are characterised by gullied glacial sediments. The distinct contrast to the bedrock slopes above defines the drift limit (Figure 13). The glacial sediments were deposited at the margin of the former Loch Lomond glacier and the drift limit, lying at approximately 200m OD (Photo 5), marks the altitude attained by the glacier on the east side of Conic Hill during the Loch Lomond Readvance. Located on the 150m contour (NS 444925) and just above the footbridge over Burn of Mar, a small, flat-topped terrace composed of sand and gravel marks the former position of an **ice-contact delta** or **kame terrace.** This was deposited at the margin of the Loch Lomond glacier during its early stages of downwasting, at the close of the Loch Lomond Stadial. The delta or kame records accumulation of sand and gravel in what may be a small, glacier-dammed lake located in the valley on the north-east side of Conic Hill. This lake may have been deeper and more extensive during the Loch Lomond Readvance maximum due to the

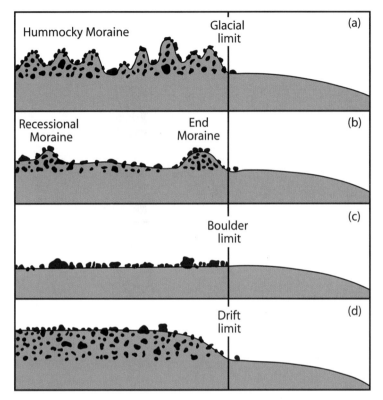

Figure 13: Various types of former glacier limits in upland terrain. When located on steep slopes, such as on Conic Hill, drift limits may become gullied by postglacial hillslope erosion.

damming of Burn of Mar but, because runoff from the surrounding slopes was not fed directly by glacier ice, no prominent depositional features were deposited in the basin except for the 150m terrace at the former glacier margin. A small but prominent **esker** runs along the hillside at NS 450924, probably documenting the sub-marginal drainage of the lake waters, and then Burn of Mar, during a slightly later stage of glacier recession. Eskers produced in this way and not fed directly by englacial and subglacial meltwater are referred to as valley eskers (also called subglacially engorged eskers), because they record the early attempts by marginal drainage systems to reoccupy their natural drainage routes through the glacier. They are typical of thin glacier margins.

Access

The West Highland Way climbs over the summit of Conic Hill and can be accessed from the B837 along farm roads at Milton of Buchanan (NS 446904). This circular route takes you over the

***Photo 5:** Loch Lomond Readvance drift limit on the southern slopes of Conic Hill viewed by telephoto from Drymen. Photo: D.J.A. Evans.*

ice-contact delta at the crossing point of the Burn of Mar and the valley esker. The drift limit on Conic Hill is very clear when approaching from this direction in late evening sunlight. A shorter walk to the summit is from the Balmaha car park (NS 422908). The view from Conic Hill takes in the islands of south Loch Lomond as well as Ben Lomond and the highlands to the north and the Campsie Fells and Kilpatrick Hills to the south.

The geology of the Highland Boundary Fault can be viewed on the Highland Boundary Fault Trail at David Marshall Lodge near Aberfoyle on the A81.

DRUMBEG AND GARTNESS

Photo 6: The Loch Lomond Readvance moraines/ice-contact deltas viewed from the A809 south of Croftamie. Photo: D.J.A. Evans.

A sand and gravel quarry at Drumbeg (NS 4887) and a river cliff at Gartness (NS 502868) afford exposures through sediments that record Dimlington and Loch Lomond Stadial events in the region. The most prominent landforms in the area are six arcuate (curved) ridges interpreted as Loch Lomond Readvance end moraines (Photo 6). The outermost ridge runs between the A811 near Lednabra (NS 501894) and Upper Gartness farm (NS 497869) and then curves south-westward towards Finnich Glen (NS 4984). The minor road that links the A811 and the farms of Drumhead (NS 496883), Gartacharn (NS 496877) and Upper Gartness (NS 496868) runs along part of the crest of this prominent moraine. Its cross-profile changes from a sharp-crest at its north end to a flat-topped ridge towards the south. In fact, between Gartacharn and the B834 near Finnich Glen the flat-topped nature of the outermost moraine ridge and its composition of Blane Valley silts indicate that it was deposited subaqueously in Lake Blane. Predominantly flat surfaces also characterise the two ridges that lie to the west, the westernmost of these containing the Drumbeg quarry.

Photo 7: Exposed sediments in Drumbeg quarry: *(a) gravel and sand delta bedding, (b) glacitectonically disturbed delta sediments and overlying deformation till, and (c) individual gravel beds deposited on the former delta front. Photos: D.J.A. Evans.*

The sedimentary structures exposed during quarrying at Drumbeg included all of the characteristics of a typical delta, including muds, sands, and gravels (Photo 7). Such deposits show the wide range of meltwater discharges that would have been released by the adjacent Lomond glacier. The construction of this delta documents the influx of sediment-laden meltwater from the margins of the Lomond glacier as it receded from its Loch Lomond Readvance maximum position (Figure 14). Such landforms are referred to as ice-contact deltas and clearly require the presence of a lake in front of the glacier snout. The altitude of the delta top at Drumbeg indicates that the lake inundated the Blane Valley up to a maximum of 67m OD (Glacial Lake Blane). The flat-topped sections of the two moraines to the east of the Drumbeg ridge suggest that they too were at least partially deposited as ice-contact deltas. Localised pushing of lake sediments by ice was probably responsible for the construction of the sharp-crested sections of these moraines.

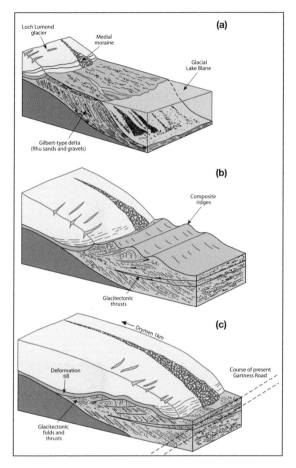

Figure 14: Reconstruction of the depositional environment and the fluctuations of the Loch Lomond glacier margin as recorded in the Drumbeg quarry sections:
the Loch Lomond glacier (a) recedes from proglacial Lake Blane, depositing an ice-contact delta, (b) readvances into the delta, producing glacitectonic folds and thrusts and stacked blocks of delta sediment in a series of composite ridges, and
(c) overrides the composite ridges, reworking the top of the delta into glacitectonically disturbed sediments and deformation till.

The uppermost deltaic sediments at Drumbeg quarry display particularly well-developed **glacitectonic** disturbance structures (Figure 15). These are similar in every way to the tectonic structures produced during mountain building events, except that the scale is several orders of magnitude smaller. Low-angle shear planes running through the upper few metres of the gravels and stacked blocks of internally folded and faulted delta sediments (Figure 15) document proglacial disturbance of the delta sequence as glacier ice readvanced over the site. Subsequent disturbance of the delta sediments by subglacial deformation is recorded in the overlying sediments. The simplified vertical sequence at Drumbeg comprises, from base to top: unit 1) undisturbed **glacilacustrine** delta sediments; unit 2) faulted and folded delta sediments; unit 3) sediments that have been squeezed out of shape, comprising sands and gravels displaying original sedimentary structures, separated by attenuated beds of clay and silt; unit 4) **deformation till**. This sequence, from undisturbed sediment, upward through folded, faulted and otherwise disturbed sediments, to a till that incorporates and transforms the underlying sediments, has been used in a classification scheme for glacial sediments by Benn and Evans (1996).

The glacitectonic disturbance at Drumbeg is also interesting from a chronological perspective. The disturbance clearly demonstrates that the delta was partially overrun after deposition by the Loch Lomond glacier (Figure 14). This constitutes a substantial readvance by the glacier (Figure 14b and 14c). The climatic significance of such readvances is often unclear. For example, a readvance of this magnitude could be driven by local conditions at the ice margin rather than by a brief period of climatic deterioration and positive mass balance.

*Figure 15: **Drumbeg quarry**: lower sketch shows details of the south section face in 1995 showing delta deposits capped by till deposited by the Loch Lomond glacier when readvancing during recession from the Loch Lomond Readvance limit to the east. The depth and intensity of glacitectonic disturbance by the glacier increases towards the west. Vertical profile shows summary of sedimentary sequence at centre of the section face, comprising four main sediment units.*

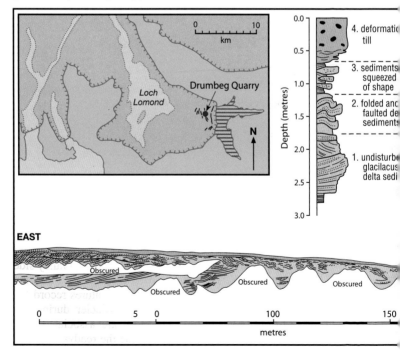

Photo 8: The meander loop in the Endrick Water at Gartness. Note the rotational landslides on the cutbank. Photo: D.J.A. Evans.

The outermost Loch Lomond Readvance moraine is dissected by Endrick Water at Gartness, where the river is characterised by a large meander loop. Although the Quaternary stratigraphy is not always clearly exposed, ongoing landslide activity on the banks of the river (Photo 8) provides occasional outcrops of the main sedimentary units, often not visible at other sites in the area. Previous exposures have shown that the sediments on the up-valley side of the moraine ridge are glacitectonised, indicating that the glacier margin disturbed them after their deposition, as envisaged at Drumbeg quarry. The Wilderness Till of the Dimlington Stadial outcrops at the base of the sequence at Gartness and is a stiff reddish-brown till with glacially modified stones. This subglacial deposit forms the drumlins that lie to the east, outside the Loch Lomond Readvance limit. Till fabrics and drumlin orientations record a former ice flow from south-west to north-east. The Wilderness Till is overlain by the Gartness laminated silts and clay, recording deposition in a proglacial, ice-dammed lake in the Endrick and Blane valleys. These sediments are overlain in turn by Clyde Beds with their characteristic remains of marine fauna. The Clyde Beds record the incursion of the sea into the Endrick and Blane valleys during the Windermere Interstadial. They are deformed on the up-valley side of the moraine and have been incorporated into the overlying Gartocharn Till. These features record the advance of the Loch Lomond Glacier during the Loch Lomond Readvance and are associated with numerous small drumlins inside the readvance limit.

The uppermost sediment unit at the site is the Blane Valley laminated silts, documenting the occurrence of an ice-dammed, proglacial lake (Lake Blane) to the east of the Loch Lomond Readvance glacier margin. The Blane Valley silts drape other glacial landforms up to an altitude of 65m beyond the readvance margin. In the area immediately to the west of the outermost moraine they are less continuous but still occur up to 65m. This indicates that the lake persisted after initial glacier recession and its upper level was controlled by the spillway at Ballat. After ice-dammed lakes drain, their former beds are often clearly identifiable due to the flat nature of the valley floor. This is a product of draping of the lake bottom with glacilacustrine silts and clays. The incision of the former lake bottom sediments by local streams grading to their new base level results in the production of gullies that cut back into the flat terrain. As tributaries run dry many gullies become abandoned, leaving a local topography very similar to coastal mudflats once the tide has gone out. This sort of topography can be viewed along the Blane Valley between Killearn House (NS 505848) and Dumgoyne where lake sediments that once blanketed the bottom of Glacial Lake Blane are responsible for the flat-lying terrain and its network of gullies (Photo 9). Towards the south, local quarrying has demonstrated that Strath Blane is composed of sand and gravel sheets deposited into Glacial Lake Blane by Blane Water as a shallow gradient fan delta.

The conspicuous **meander** (NS 4986) in Endrick Water at Gartness is significant with respect to postglacial drainage evolution. It lies over the axis of a buried valley that was occupied by Endrick Water before the Dimlington Stadial glaciation. The river was forced to occupy a more easterly route between Gartness Mill (NS 497861) and its junction with Blane Water (NS 504855) by the plug of Dimlington

Photo 9: The former lake bottom of Glacial Lake Blane near Dumgoyne. *Note the dry gullies that were incised into the lake sediments by local streams after the lake drained. Photo: D.J.A. Evans.*

Stadial sediments and the outermost Loch Lomond Readvance moraine blocking the old bedrock course between Gartness and Dalnair (NS 497861). The old bedrock valley side is exposed at Gartness Mill and is the location of the present waterfall on the river. The meander that has developed upstream of the waterfall is the product of erosion in the unconsolidated Quaternary sediments blocking the old river course. As the altitude of the river is some 13m higher on the east side of the moraine than on the west side, it will not be long on the geological timescale before Endrick Water breaches the blockage to reoccupy its old course and abandons its meandering reach between Gartness Mill and the junction with Blane Water.

Access

An excellent vantage point to view the south Loch Lomond basin and the glacial features of the Loch Lomond Readvance is at the 70m triangulation pillar at Drumbeg quarry (NS 485879). The Loch Lomond Readvance drift limit on Conic Hill can be seen to the north-west, drumlins fill the middle distance to the south-west, and the Loch Lomond Readvance end moraines fill the view to the east.

The Gartness landslides occur at the meander neck in Endrick Water at NS 498865. Exposures in the Quaternary stratigraphy are not always clean and depend upon recent erosion rates. The landslides are on private land and permission should be sought before accessing the site. At the time of writing (August 2003) the Drumbeg quarry was disused and future landscaping was uncertain. Good exposures in the glacitectonised sediments were still available at the south-west corner of the quarry but access permission must be sought from the owner via the employees at the stables at the north end of the quarry (NS 478882).

The former bed of glacial Lake Blane can be viewed from the A81 between Killearn House and Dumgoyne, although the best vantage point is from Dumgoyne Hill (NS 542828), a short climb that begins at the Glengoyne distillery (NS 526827).

CROFTAMIE

An exposure through Quaternary sediments in the abandoned railway cutting (NS 473861) at Croftamie is the official type site of the Loch Lomond Readvance/Stadial in its type area (Figure 6). This status is justified by the fact that it is the only locality in Britain where plant remains separate the glacial deposits of the Dimlington and Loch Lomond stadials. The exposure is located just below the crestline of the local Loch Lomond Readvance drumlins. At the base of the exposure the Dimlington Stadial Wilderness Till lies directly on the local Old Red Sandstone bedrock. The till contains a lot of locally derived stones that are relatively angular compared to most subglacial deposits. This indicates that the glacier was plucking the local bedrock and mixing the freshly quarried fragments with a matrix of sands and silts to produce a deformation till. Till fabrics indicate a north-west to south-east ice flow direction. Overlying the Wilderness Till is a layer of organic detritus containing pollen grains indicative of frost-disturbed ground and typical of vegetation dating to the Devensian Lateglacial. A reindeer antler was also collected from the organics in the 1850s. The organic detritus is overlain by the Blane Valley laminated silts and clays, deposited in the proglacial lake that was formed as the Loch Lomond Glacier moved into the lower Endrick Valley. The arrival of the glacier at the site is recorded in the

Photo 10: Drumlins molded by the Loch Lomond glacier *during the Loch Lomond Readvance at Croftamie. Photo: D.J.A. Evans.*

Photo 11: The Croftamie exposure through Quaternary sediment, after excavation by mechanical digger. Photo: D.J.A. Evans.

exposure by the erosional truncation of the lake sediment and the overlying Gartocharn Till. The till fabric in the Gartocharn Till records a former ice flow from north-west to south-east, parallel with the local drumlins (Photo 10). Fragments of marine fauna in the till were derived from the Clyde Beds at locations to the north of Croftamie. As the recent exposure (Photo 11) shows red coloured sands at the base lie on top of Wilderness Till. The sands are part of the 'Gartness gravel, sands and laminated silts and clay' deposited in Glacial Lake Blane during the retreat of the Dimlington Stadial ice sheet. These sands are overlain by a black coloured organic detritus layer, radiocarbon dated at 10,560±160, documenting vegetation growth during the Windermere Interstadial. The grey layer overlying the organic detritus in Photo 11 is the Blane Valley laminated silts and clays, deposited in Glacial Lake Blane as the Loch Lomond glacier advanced into the area during the Loch Lomond Stadial. The Gartocharn Till, deposited by the glacier, is the uppermost reddish-brown layer.

Access

The Quaternary stratigraphy is protected as a site of special scientific interest and as such is available for any visitors to access. It crops out in the embankment of the disused railway line (NS 473861) but requires a considerable amount of cleaning before any clear stratigraphic details can be observed. The Croftamie drumlins are best viewed from the lay-by on the A809 (NS 475867). Further excellent examples relating to the Loch Lomond Readvance occur near Buchanan Smithy (NS 466896) on the B837 immediately north of Drymen. They can all be viewed from various angles from local roads.

CAMERON MUIR, CARNOCK BURN AND FINNICH GLEN

The southern margin of the Loch Lomond Glacier during the Loch Lomond Readvance moved uphill on the northern slopes of Cameron Muir where it deposited a prominent end moraine. This moraine can be viewed in the vicinity of Carnock Burn, to the southwest of Aucheneck House, specifically in the area between NS 476827 and NS 480828. It is mostly a single ridge feature up to 7m high and 40m wide on Cameron Muir but does comprise multiple ridges towards the western part of the area. To the north of the moraine the landscape is characterised by numerous irregularly spaced till hummocks that were most likely deposited on the surface of the glacier as it downwasted. In contrast, the landscape (supraglacial hummocky moraine) to the south contains few till hummocks and these are more subdued due to the fact that they were deposited by downwasting Dimlington Stadial ice and therefore were subjected to periglacial processes during the Loch Lomond Stadial.

The moraine descends into Carnock Burn causing a small river diversion at NS 476827, thereby demonstrating that the valley existed prior to the Loch Lomond Readvance. Small exposures in the moraine in the vicinity of NS 482837 reveal that it is composed of dark reddish brown sandy till. The till is seen to thicken along Carnock Burn wherever it infills the pre-existing rock-cut valley of the Burn. There is a striking contrast in the form of the valley on either side of the moraine.

To the south the valley is relatively broad, the **floodplain** is continuous and the valley sides are composed of the Old Red Sandstone bedrock. To the north of the moraine, the valley floor is no wider than the stream and the valley sides are often composed of Loch Lomond Readvance till. From time to time, the stream occupies a bedrock gorge cut in the floor of the valley. This gorge is best seen at Finnich Glen (NS 4984), a remarkable deeply incised slot gorge that runs beneath the A809. The filling of the wider bedrock valley with till inside the Loch Lomond Readvance moraine and the lack of valley fill to the south of the moraine indicates that a large valley was produced at some time prior to the readvance. This was probably excavated by meltwater from the wasting Dimlington Stadial ice sheet. The meltwater could have drained subglacially or marginally or by a combination of these pathways. The rock-cut valley was then partially plugged with till during the Loch Lomond Readvance. Since the Loch Lomond Readvance, Carnock Burn has incised down through the till plug and reoccupied the base of the rock-cut valley but failed to clear out all the till from the wider buried valley, explaining

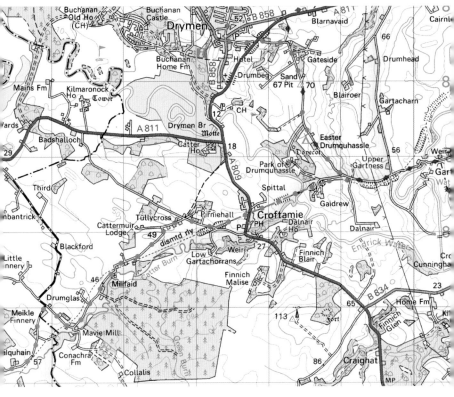

© *Crown Copyright*

the contrast in valley form on the two sides of moraine. The fact that Finnich Glen has been incised to a depth lower (*ca.* 30m OD) than the upper level of Glacial Lake Blane indicates that bedrock incision by Carnock Burn continued after the lake was drained.

Access and safety

Each of these sites is best visited on foot from the large parking area located at the junction of the A809 and the B834 at NS 493850. To reach the sites on Carnock Burn take the track westwards. Finnich Glen was a popular place for well-informed locals seeking a short walk or attempting something more daring by swimming/scrambling along the gorge. However, in early 2001 a gate was erected by the landowner with a warning to keep out of the east side of the glen due to the risks. **Finnich Glen is a potentially dangerous place to be. Beyond the gate there is easy access to the precipitous drop into the gorge and the upper grassy slopes can be slippery when wet. All parties wishing to visit the gorge must be fully insured and must seek permission from Mr D. Young at Killearn Home Farm.**

THE WHANGIE

The Whangie to the north-west summit of Achineden Hill is a well-known destination for a short walk along the north-facing scarp of the Kilpatrick Hills. The term literally means 'cut' or 'slice' and refers to the spectacular 10m-deep chasms separating slabs of basalt from the cliff face (Photo 12). The chasms are up to 2m wide at the base and the complex cleavage planes on both walls would fit together like a jigsaw puzzle if the blocks were pushed back against the cliff. Legend credits the cutting of the main chasm to the devil when he lashed the plateau edge with his tail whilst flying over the area. The feature is almost certainly a translational landslide or block guide failure (Figure 12, page 24). One 10m high and 3m wide block remains standing and older blocks in various stages of collapse lie downslope.

The details of the sub-surface bedrock characteristics are not well-known but it appears that the blocks of basalt are gliding over a failure plane that may lie at the junction with underlying sandstones.

Photo 12: The Whangie from the south: *(a) a large slab of basalt detached from the main rock face, and (b) the detached slab and chasm. Photos: D.J.A. Evans.*

This may arise from the underlying rocks being less competent than the basalt or having a radically different permeability. Either scenario allows the development of a low-angled slide plane over which the basalt blocks move before being fractured by tensional stresses. This fracturing produces slabs whose size and shape are dictated by the structural weaknesses such as faults and joints in the basalt. As an individual slab moves downslope it becomes unstable and topples over, thereby fracturing and producing a rubble veneer. The exact age of the Whangie is unknown but, like the Campsie Fells landslides, it may have been produced during the early stages of the paraglacial cycle or could have been initiated by earthquake activity.

Access

The Whangie is the highlight of a popular short walk from the Auchineden (Queen's View) car park on the A809 at NS 511807. The footpaths are clearly marked and the walk is not strenuous. Access can be gained to the chasms between individual slide blocks at NS 492806

THE ENDRICK WATER FLOODPLAIN

Where Endrick Water enters the south-east corner of Loch Lomond it has deposited a large volume of sediment on a wide **floodplain**. Much of this sediment is derived from the Endrick raised delta, originally deposited at 12m OD when the sea inundated Loch Lomond between 6900 and 5450 years BP and now incised by the meanders of lower Endrick Water. It is over the floodplain and through the delta that the river now meanders between Croftamie and the loch shore, providing excellent exposures through the fine-grained alluvium in erosional bluffs and good examples of **cut-off meanders, point bars** and **bar and swale topography.** Modern point bars are exposed during periods of low flow.

Endrick Water is worth visiting even during inclement weather, because the river will often flood the fields on either side of the A809, providing a spectacular illustration of why humans should think twice

Photo 13: Aerial view over southern Loch Lomond: the simple birds-foot delta that has developed at the point where the Endrick Water enters the loch is clearly visible. Photo: British Geological Survey (www.bgs.ac.uk). © NERC. All rights reserved.

about constructing buildings on floodplains! The volume of sediment transported by Endrick Water is immediately apparent when viewing the mouth of the river from the air (Photo 13). From this vantage point you can see that the river has constructed a 2km-long bank along the south-east shore of the loch, in the form of a simple **birds-foot delta.**

Access

The river geomorphology of Endrick Water can be accessed from the Drymen Bridge on the A809 at NS 473874 where a small lay-by provides parking space. There is open access, by kind permission of the landowner, to the fields immediately to the west of the bridge.

ROWARDENNAN PIER

Although the impact of glacial erosion is most impressive at the larger scale in this area (e.g. Loch Lomond itself), some excellent examples of micro scales of **abrasion** by glacier ice are to be found on the shores of the Loch. At Rowardennan Pier, for example, the Precambrian bedrock has been smoothed into **whalebacks** by the debris carried in the base of the overriding Loch Lomond glacier (Photo 14). Superimposed on the whalebacks are smaller abrasion features such as **P-forms** (in this case smooth-sided channels) and **striae**. P-forms have always been a matter of contention in glacial geomorphology, because researchers cannot agree on their exact origin. Because they look like normal fluvially incised channels, but go up and down slopes and possess abrupt corners, some researchers have suggested a subglacial fluvial origin whereby bedload in

Photo 14: Whalebacks at Rowardennan pier. Ice flow was down the loch from left to right. Photo: D.J.A. Evans.

pressurised streams scours the P-forms. However, some P-forms are covered in striae which often follow their surface contours, indicating that debris-charged glacier ice flowed along and abraded the feature. Certainly a combination of the two processes is possible. The debris-charged basal layers of modern glaciers have been observed in subglacial cavities flowing in a variety of directions all at the same time. Therefore, a striated surface recording a variety of ice flow directions may have been abraded during just one glacier overriding event.

Access

The glacial erosion features lie on the banks of Loch Lomond on the edge of the Rowardennan car park at NS 359988. This is situated at the end of the minor road that continues on from the B837 at Balmaha. Numerous other examples exist on the West Highland Way, which continues from the car park northwards along the loch shore.

WESTERN FORTH VALLEY

The moraine of the Loch Lomond Readvance Menteith glacier in the western Forth Valley comprises a spectacular belt of multiple ridges and intervening depressions, forming an arc around the southern boundary of west Flanders Moss and continuing north to south from Arnprior to the Lake of Menteith (Figure 16). Moraine units A and C on Figure 16 are composed of thick sequences of glacitectonically thrust Quaternary sediments with ridge crests lying across the main valley. Moraine unit B comprises an area of low amplitude, variously aligned sand and gravel ridges lying on a bedrock high point in the centre of the main valley. Moraine unit D is the easterly extension of the prominent dual moraine ridges that extend from the hillside north of Garrauld. The ridges and depressions are most pronounced between Buchlyvie and Arnprior, where several meltwater channels have been incised through the moraines, and on the eastern shore of the Lake of Menteith. Between these two areas the moraine is characterised by small sand and gravel ridges draping a bedrock high. It is possible that these ridges are former crevasse-fill sediments. Several characteristics of the large moraine on the east shore of the Lake of Menteith suggest that it is a **hill-hole pair**, a glacitectonic landform (Figure 17). Hill-hole pairs are represented by a hill of glacier-thrust material located a short distance down glacier from a depression of similar size and volume. The hill is essentially a thrust and compressed mass of sediment that has been removed from its original location by glacier overriding, leaving the hole as a source depression. Because they are often overridden and modified by glacier ice and the depressions may become infilled by lake sediments and peat, hill-hole pairs are difficult to identify. Many more pairs may be located in glaciated landscapes like south Loch Lomond and the western Forth Valley.

The moraine to the east of Lake of Menteith lies immediately down valley from a large, water-filled depression (Lake of Menteith) from which it was excavated by glacier ice, and the moraine surface is composed of numerous parallel ridges representing individual thrust slices or fold noses (Photo 15). Such ridges are often gradually smoothed out when glaciers continue advancing and override them, but this appears not to have been the case at the Lake of Menteith.

Because the glacier that excavated the hill-hole pair removed pre-existing sediments from the Lake of Menteith depression, the thrust hill must contain evidence of those sediments. This was verified by early researchers, who discovered sections near Inchie (NS 592000) that contained marine clays overlain by sand and gravel.

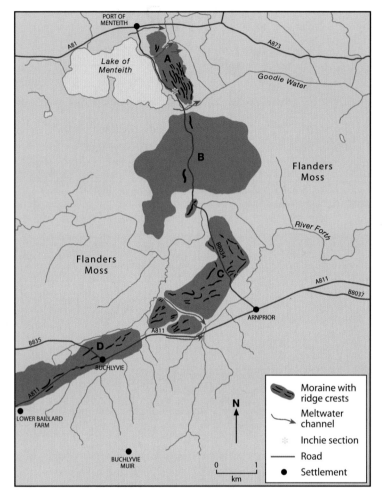

*Figure 16: **Moraine belt and associated landforms** produced at the former margin of the Loch Lomond Stadial Menteith Glacier.*

The marine clays contain fragments of the marine mollusc *Mytilus edulis* (the common mussel) which have been radiocarbon dated to 11,800±170 years BP, confirming that the deposits were laid down prior to the Loch Lomond Readvance and that a glacier dating to the readvance constructed the Menteith moraine.

The moraine belt between Arnprior and Buchlyvie may also be a hill-hole pair or a composite ridge, the latter being a slightly different type of glacitectonic landform (Figure 17). The source depression lies under the peat of west Flanders Moss. Viewed from above, a composite ridge comprises an arcuate series of subparallel ridges and intervening depressions, arranged largely parallel to the glacier snout (Photo 16b). There is a close association between composite ridges

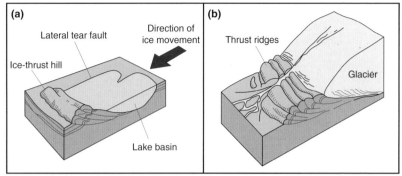

Figure 17: (a) A typical hill-hole pair after glacier recession, and (b) A typical composite ridge. Source: Evans and Benn, 2001.

Photo 15: Aerial orthophotograph of the hill-hole pair at the Lake of Menteith, showing the parallel, cross-valley ridges produced by folding and thrusting of sediment slices by the Menteith Glacier during the Loch Lomond Readvance. Compiled by S.B. Wilson, University of Glasgow based on RCAHMS, Edinburgh photo number 51288128 – Crown Copyright.

and two types of glacier snout: (1) surging glaciers in Iceland and Svalbard where the snout advances and applies stress on proglacial sediments rapidly; (2) sub-polar and polar glaciers like the Canadian Arctic (Photo 16a) and Antarctica where the glacier has very slowly dislocated the proglacial frozen ground (permafrost), probably due to failure in unfrozen sediment bodies (taliks) within the permafrost. It may not be possible to determine exactly what sort of glacier snout conditions prevailed in the western Forth Valley at the time of the Loch Lomond Readvance, but the occurrence of crevasse-fill ridges in association with hill-hole pairs and composite ridges is typical of many present-day surging glacier snouts.

The southern limit of the Menteith Glacier during the Loch Lomond Readvance is demarcated by a prominent dual moraine ridge that extends from Buchlyvie to the hillside north of Garrauld. A large meltwater channel runs along the outside of the moraine, parallel to the A811 between the farms of Kepculloch and Lower Ballaird. A small delta at the northern end of the Ballat spillway lies immediately outside the readvance limit of the Menteith Glacier. The delta records the damming of a small lake at the ice margin, although the lake water was probably draining under or through the glacier in order to maintain the level of Glacial Lake Blane at 65m. The land outside the Loch Lomond Readvance limit consists of south-west to north-east trending drumlins, documenting the flow of Dimlington Stadial ice along the northern slopes of Buchlyvie Muir.

The extensive and remarkably flat agricultural land of the Forth Valley (Figure 18 and Photo 17), locally named Flanders Moss, is a product of estuarine deposition of clays and occasional sands, referred to locally as **carse clays** or **carselands,** during the period of relatively high (glacioeustatically controlled) sea level between 6900 and 5450

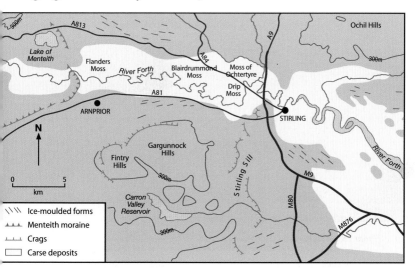

Figure 18: The extent of the carse deposits in the Forth valley.
After: Hansom and Evans, 2000.

— CLASSIC LANDFORMS —

***Photo 16:** (a) **A modern composite ridge at the margin of a glacier in the Canadian Arctic,** with (b) a view of the the glacitectonic thrust ridges at Buchlyvie. Photos: D.J.A. Evans.*

Photo 17: View across the carselands in the upper Forth Valley. Photo: J.D. Hansom.

years BP. The estuarine sediments are now raised above present sea level due to the glacioisostatic rebound of the crust since their deposition. The margins of the carselands give way abruptly to the steep slopes of the Forth Valley and the Loch Lomond Readvance landforms of the former Menteith Glacier. Because the sea was able to penetrate behind the Menteith Glacier end moraine at the end of the Loch Lomond Stadial, carse clays continue west of the composite ridges at Arnprior (Figure 18). Interesting finds within the carse clays include a whale skeleton at Blairdrummond Moss. This was found in association with human hunting implements, indicating that people had been hunting whales at a time when the area lay below the sea.

Access

Excellent views of the moraines and meltwater channels to the south of Flanders Moss are available from the A811 between Kepculloch (NS 542918) and Arnprior. The lateral moraine of the Menteith glacier trends northwestwards from Kepculloch along the north side of Moor Park. Individual thrust blocks within the composite ridge are located close to the lay-by on the east side of Buchlyvie. Parking and picnic facilities are available on the B8034 where it skirts the east end of the Lake of Menteith and follows the base of the ice-thrust hill (NS 612948).

GLOSSARY

Abrasion. The scoring of bedrock by debris carried in the base of a glacier, producing striae (see Striation below).

Bar and swale topography. That part of a floodplain characterised by a series of bars and intervening troughs (swales). As each bar represents an individual point bar, the full series records migration of the river away from a slip-off slope during meandering.

Birds-foot delta. A delta formed by the growth of river levees into a deep-water body to produce a series of finger-like depositional ridges. The number of 'fingers' reflects the number of distributary channels in the river.

Carse clays/carseland. Flat-lying areas of former estuarine muds now raised above sea level due to glacioisostatic rebound.

Cirque. Amphitheatre-shaped bedrock hollow composed of a precipitous backwall and overdeepened floor. Backwall recession is thought to be a product of frost shattering in the *bergschrund* of the glacier. Floor overdeepening is a product of subglacial erosion by the glacier occupying the cirque basin.

Deformation till. Rock or sediment that has been disaggregated and completely or largely homogenised by shearing in a subglacial deforming layer. Most of the apparently massive or structureless tills in glaciated landscapes are now thought to be the product of disaggregation and homogenisation of pre-existing materials.

Drumlin. A glacially streamlined hill composed of till and/or other unconsolidated materials.

Erratic. A piece of rock that has been carried by glacier ice and deposited in an area remote from its place of origin.

Esker. A long, narrow ridge of sand and gravel, originally deposited in a tunnel within or beneath glacial ice.

Floodplain. The part of the river valley adjacent to the channel over which the river flows during floods. Rivers will rework alluvium on the floodplain on a cyclical basis due to meandering and the incision of new channels.

Glacilacustrine. Relating to lakes dammed by glacier ice or fed directly by glacial meltwater.

Glacioeustatic sea level change. The rise or fall in global sea level controlled by the decay or build-up respectively of large ice sheets. Essentially, global sea level falls when ice sheets grow and rises when ice sheets melt and release large volumes of water to the global oceans. In regions affected by glacioisostasy, such as Scotland, the glacioeustatic sea level trend is superimposed on the glacioisostatic rebound and can cause a sea level rise only if it exceeds the rate of rebound. This was accomplished between 8000 and 5450 years BP when the North American and Scandinavian ice sheets were melting but the British ice sheet had disappeared and glacioisostatic rebound was almost complete and slowing down.

Glacitectonics. The disruption of sediment or bedrock by an overriding glacier, manifest in structures such as shear faults and folds and landforms such as thrust moraines.

Gyttja. A nutrient-rich peat or organic mud which contains plankton.

Hogsback ridge. A ridge formed by steeply dipping strata so that the slopes on both sides of the feature are steep.

Ice-contact delta. A delta that has been built out into a lake or the sea by the meltwater from a glacier snout. Unlike a normal delta, an ice-contact delta possesses a steep, former ice-contact face where the glacier once abutted the feature.

Kame and kame terrace. Mound and flat-topped linear form resulting from the deposition of stratified sediments at the glacier margin by glacial meltwater.

Meander. A loop-like bend in a sinuous river channel. An erosional bluff exists on the outside of the meander curve and a gently shelving, depositional slip-off slope occurs on the inside. A cut-off meander is formed where the river cuts across and abandons a loop. Cut-off meanders are often occupied by ox-bow lakes.

Periglacial. Relating to intensely cold environmental conditions adjacent to glaciers or, more generally, in high latitudes or at high altitudes.

P-form. A glacially smoothed groove. They come in a variety of shapes and sizes ranging from straight and sinuous grooves to crescentic-shaped scours called *sichelwanne* (sickle shape).

Point bar. An elongate depositional feature composed of sand and gravel and located on the slip-off slope of a river meander. It is separated from the river bank by an elongate depression or swale.

Radiocarbon date. An age assigned to organic remains based upon the remaining carbon 14 (C14), which breaks down after the death of the organism. Throughout this guide, the ages provided are given in radiocarbon years, but recent advances in the technique have highlighted the fact that radiocarbon years do not exactly equate to calender years. Therefore, radiocarbon dates are now being expressed in calibrated radiocarbon years so that, for example, the calibrated age range for the Loch Lomond Stadial is 12,800-11,500 years BP compared to the radiocarbon age of 11,000-10,000 years BP.

Striation (plural striae). A scratch on a rock surface produced by the process of abrasion. This occurs when stones protruding from the base of a sliding glacier are dragged across a rock surface.

Till/Till fabric. Till is an unstratified, poorly sorted sediment deposited directly by glacier action. Till fabric is the alignment of stones within the till which can be measured to determine the direction of ice movement.

Volcanic plug. The remnants of a former volcanic neck, represented by the solidified contents of the central vent once the surrounding cone or lavas have been removed by erosion.

Whaleback. A glacially abraded and polished rock surface that looks like the back of a whale breaking the surface of the ocean.

BIBLIOGRAPHY

Ballantyne, C.K. (1986) 'Landslides and slope failures in Scotland: a review', *Scottish Geographical Magazine,* 102, pp. 134-50.

Benn, D.I. and Evans, D.J.A. (1996) 'The interpretation and classification of subglacially deformed materials', *Quaternary Science Reviews,* 15, 23-52.

Benn, D.I. and Evans, D.J.A. (1998) *Glaciers and Glaciation.* London: Arnold.

Boulton G.S., Peacock J.D. and Sutherland D.G. (1991) 'Quaternary' in Craig, G.Y. (ed) *Geology of Scotland.* London: The Geological Society, pp. 503-43.

Browne, M.A.E. and Mendum, J. (undated) *Loch Lomond to Stirling: A landscape fashioned by Geology.* Edinburgh: British Geological Survey and Scottish Natural Heritage.

Evans, D.J.A. (1991) 'Glaciated land on the rebound', *Geography Review,* 4, pp. 2-6.

Evans, D.J.A. (ed) (2003) *The Quaternary of the Western Highland Boundary – Field guide.* London: Quaternary Research Association.

Evans, D.J.A. and Benn, D.I. (1998) 'The earth movers', *Geography Review,* 11, pp. 27-33.

Evans, D.J.A. and Benn, D.I. (1999) 'Glaciers as streamlining agents', *Geography Review,* 12, pp. 2-5.

Evans, D.J.A and Benn, D.I. (2001) 'Earth's giant bulldozers', *Geography Review,* 14, pp. 22-33.

Evans, D.J.A. and Hansom, J.D. (1998) 'Scottish Landform Examples 18: The Whangie and the landslides of the Campsie Fells', *Scottish Geographical Magazine,* 114, pp. 192-6.

Gordon, J.E. and Sutherland, D.G. (eds) (1993) *Quaternary of Scotland.* London: Chapman & Hall.

Hansom, J.D. and Evans, D.J.A. (2000) 'Scottish Landform Examples 23: the carse of Stirling', *Scottish Geographical Journal,* 116, pp. 71-8.

Jardine, W.G. (ed) (1980) *Glasgow Region Field Guide.* Glasgow: Quaternary Research Association.

Linton, D.L. and Moisley, H.A. (1960) 'The origin of Loch Lomond', *Scottish Geographical Magazine,* 76, pp. 26-37.

McKirdy, A. and Crofts, R. (1999) *Scotland – The Creation of its Natural Landscape: A landscape fashioned by geology.* Edinburgh: Scottish Natural Heritage/British Geological Survey.

Pierce, L. (1999) 'Loch Lomond: an example of Quaternary megageomorphology', *Scottish Geographical Journal,* 115, pp. 71-80.

Price, R.J. (1983) *Scotland's Environment During the Last 30,000 Years.* Edinburgh: Scottish Academic Press.

Rose, J. (1980) 'Gartness' in Jardine, W.G. (ed) *Glasgow Region Field Guide.* Glasgow: Quaternary Research Association, pp. 46-9.

Rose, J. (1981) 'Field guide to the Quaternary geology of the south-eastern part of the Loch Lomond basin', *Proceedings of the Geological Society of Glasgow 1980-81,* pp. 1-19.

Rose, J. (1989) 'Stadial type sections in the British Quaternary' in Rose, J. and Schluchter, C. (eds) *Quaternary Type Sections: Imagination or reality?* Rotterdam: Balkema, pp. 45-67.

Sissons, J.B. (1967) *The Evolution of Scotland's Scenery.* Edinburgh: Oliver & Boyd.

Thorp, P.W. (1991) 'Surface profiles and basal shear stresses of outlet glaciers from a Lateglacial mountain icefield in western Scotland', *Journal of Glaciology,* 37, pp. 77-89.